RUGGED RAX

The Little Satellite That Could

WRITTEN BY
Suzanne Jacobs Lipshaw

ILLUSTRATED BY
Mesa Schumacher

To Josh, my space enthusiast and to Jeremy, my sunshine.

—SUZANNE JACOBS LIPSHAW

For Zephyr and Dune—keep expanding like the universe.

—MESA SCHUMACHER

Fifth Avenue Press is a locally focused and publicly owned publishing imprint of the Ann Arbor District Library. It is dedicated to supporting the local writing community by promoting the production of original fiction, non-fiction, and poetry written for children, teens, and adults.

Printed in the United States of America

First Printing 2025

ISBN: 978-1-956697-49-0 (Hardcover)

Fifth Avenue Press
343 S. Fifth Avenue
Ann Arbor, MI 48104
fifthavenue.press

EDITOR
Dr. Virginia Loh-Hagen

ILLUSTRATION
Mesa Schumacher

LAYOUT & DESIGN
Ann Arbor District Library

Author's Note

My son, Josh, while a junior at the University of Michigan, introduced me to CubeSats when he joined the Michigan eXploration Laboratory (MXL) in 2013. During his two years there, the MXL team designed, built, and tested their fourth, fifth, and sixth CubeSats: M-CUBED-2, GRIFEX, and CADRE. M-CUBED-2 and GRIFEX had successful missions, but CADRE had a failure. CADRE's antenna became trapped underneath a solar panel that didn't move to its correct position. This flaw meant the satellite couldn't communicate with the lab.

Failure is difficult, but dedicated scientists and engineers know failure is an important part of their process. Most new discoveries are made by making mistakes, analyzing errors, and applying lessons learned on future projects.

When Josh graduated, he took the lessons he learned from both successful and failed MXL missions with him to Lockheed Martin Space where he is a space systems engineer. At Lockheed Martin, Josh was part of a team that designed and launched several GPS satellites. Currently, he is working on sending the next generation of astronauts back to the moon.

By designing spacecraft that explore Earth and beyond, engineers and scientists like Josh help us learn more about our planet and continually invent new products that enrich our lives. My hope is that the story of the engineering team that created RAX will inspire you to become a problem solver. From there, you can use your skills to positively affect our lives on Earth and, maybe even one day, on a planet far, far away.

—Suzanne Jacobs Lipshaw

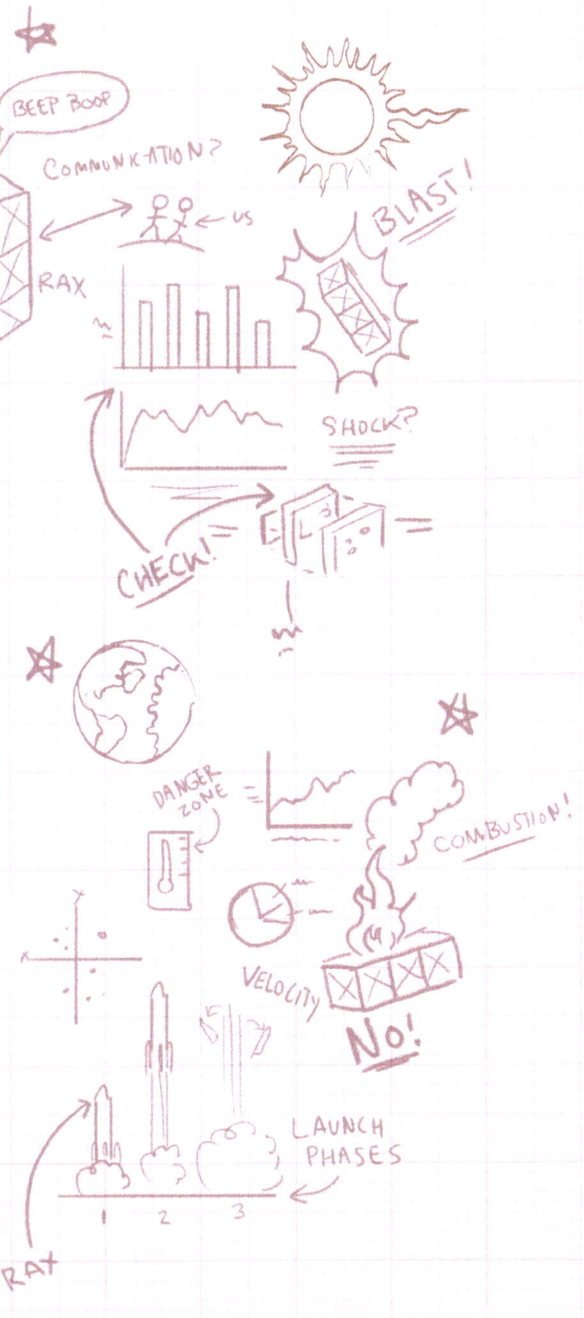

Look to the sky. A storm cloud rages.

Rain splashes. Lightning flashes.

Thunder roars.

Wind soars.

Lights flicker. TVs fizzle. The internet flops.
Streets go dark.

On the sun, a different storm rages.

Superheated gases crash.

Spurts of energy flash.

Solar winds roar.

Some soar a million miles per hour...

...all the way to Earth!

Darkness covers the city, state, and country. It even covers the continent. People wonder—what happened?

Our Sun is a fiery ball of superheated gases. The center, or core, of the sun is the hottest part. It can reach temperatures as high as 28 million degrees Fahrenheit (about 15 million degrees Celsius).

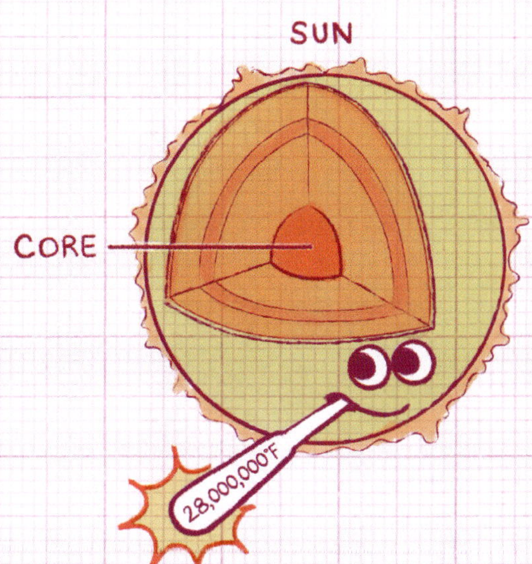

In the core, hydrogen atoms are jammed together. This pressure and the extreme heat turn the hydrogen atoms into energetic helium atoms. The helium has so much energy, it bursts away from the sun creating solar wind.

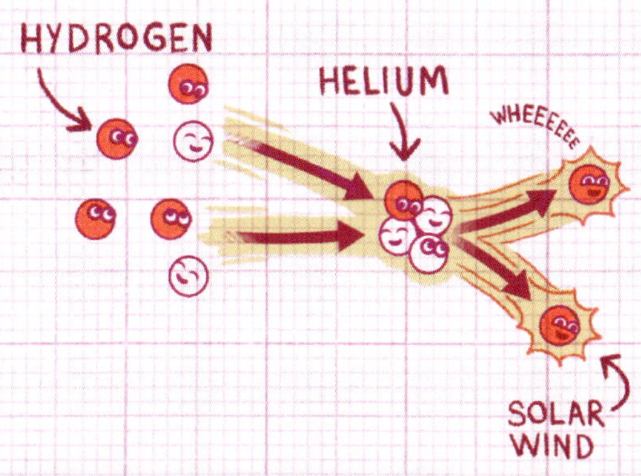

Space weather happened. Solar storms are rare.
But they happen. People need to be prepared.

Around Halloween night 2003, the sun played a ghoulish trick. It sent a monster solar storm hurtling toward Earth.

The storm shut down orbiting satellites that control electronics. TVs went dark. Cell phones stopped working. Spooked pilots couldn't land.

Luckily, the satellites came back online quickly and there was little damage.

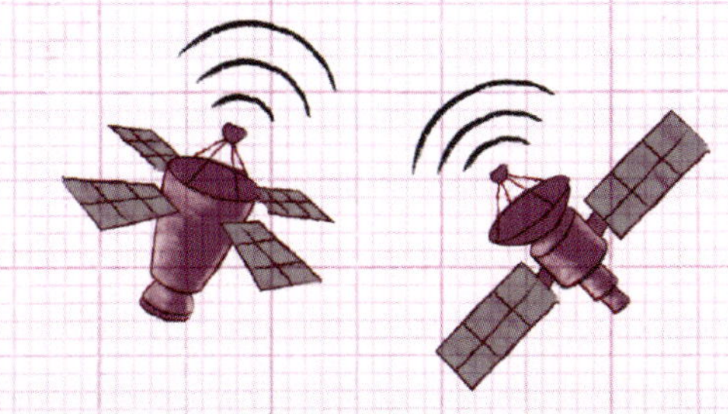

But forecasting space weather is tricky.

Enter a team of student engineers. Their mission? To create a small spacecraft to circle Earth and gather space weather data.

What if you were part of that team?

Awesome! But...

...this is the same team whose first space weather spacecraft failed.

Will they learn from their mistakes?

The real-life team tasked with creating a small space weather spacecraft were college students from the University of Michigan's eXploration Lab (MXL). They worked long and hard to design, build, and launch this spacecraft. But, as often happens on first tries, something went wrong. The spacecraft's solar panels stopped working, causing its power system to break down. The mission failed, but the team refused to give up.

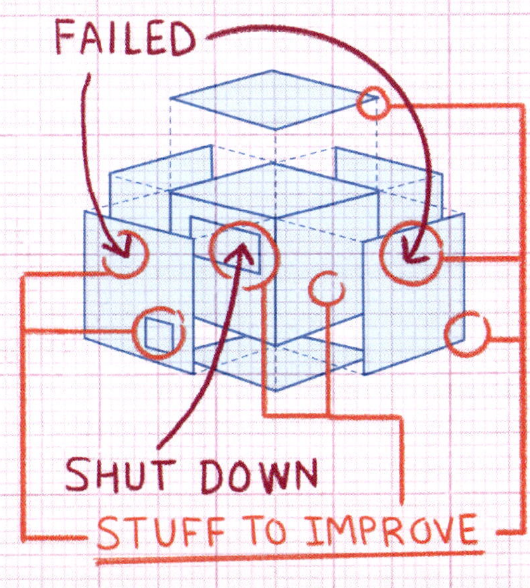

The team pushes up their sleeves. Again, they plan. They problem solve. They design another mini but mighty satellite—another CubeSat.

The team starts with a metal frame. They shape it into a cube small enough to sit in the palm of one hand. But how can they squeeze in all the parts their satellite needs to run? *Can a CubeSat be another shape?*

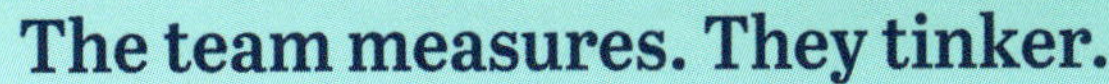

The team measures. They tinker.

They measure some more. Another metal frame is built. It's shaped into a rectangle the size of a loaf of bread.

Hooray, everything fits! The team names their CubeSat RAX.

Even though RAX is a rectangle, it is still a CubeSat because cubes are its basic building blocks. A one-unit CubeSat is 10 x 10 x 10cm^3, about the size of a softball but square. RAX is three of these cubes. Compare that to a communications satellite, which is the size of a small school bus. The International Space Station is the size of a football field.

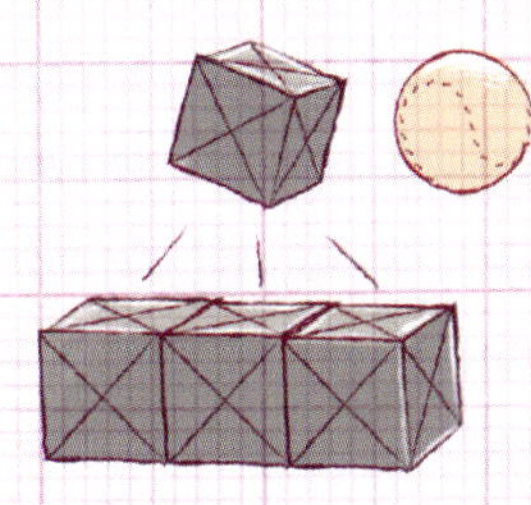

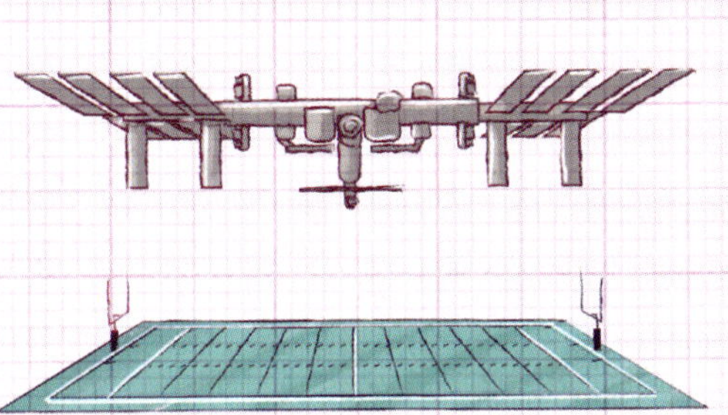

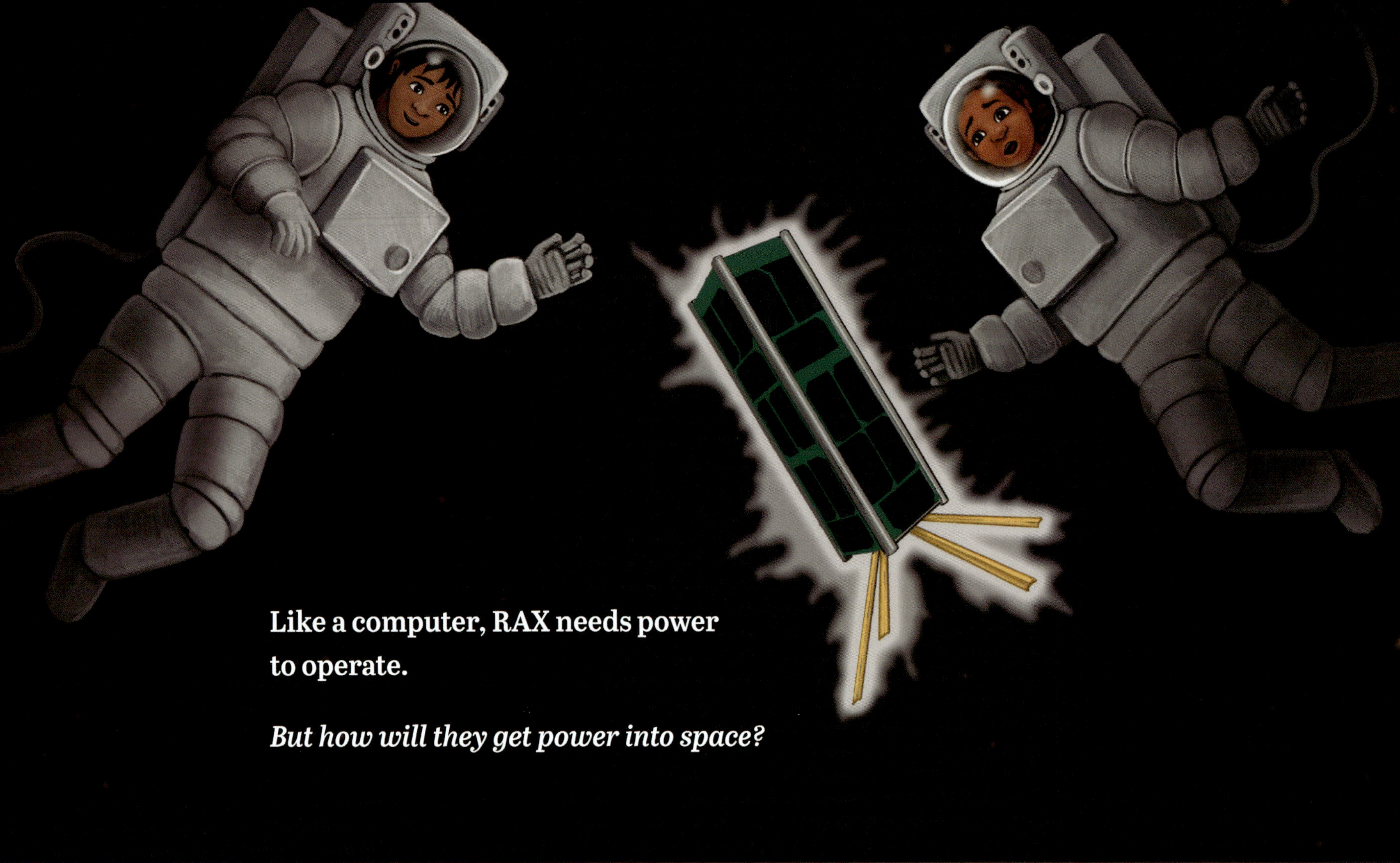

Like a computer, RAX needs power to operate.

But how will they get power into space?

Space weather isn't all bad. During a solar storm, energy particles from the sun interact with gases in Earth's atmosphere.

These gases sometimes create colorful ribbons of lights called auroras. RAX, short for Radio Aurora Explorer, is named after these dazzling light shows.

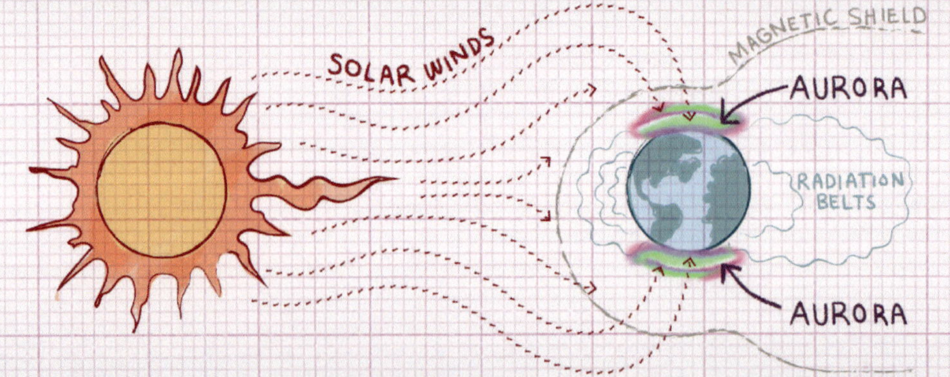

An extension cord?

No. There isn't one long enough to reach outer space.

A cell phone charger?

Nope. That would need an extension cord too.

Gasoline?

No go. There are no gas pumps in space.

The sun?

Yes!

The sun is a ball of energy. With energy comes power.

The team fastens solar panels to RAX. This will soak up power from the sun.

Next, RAX needs to talk to the team. *But how?*

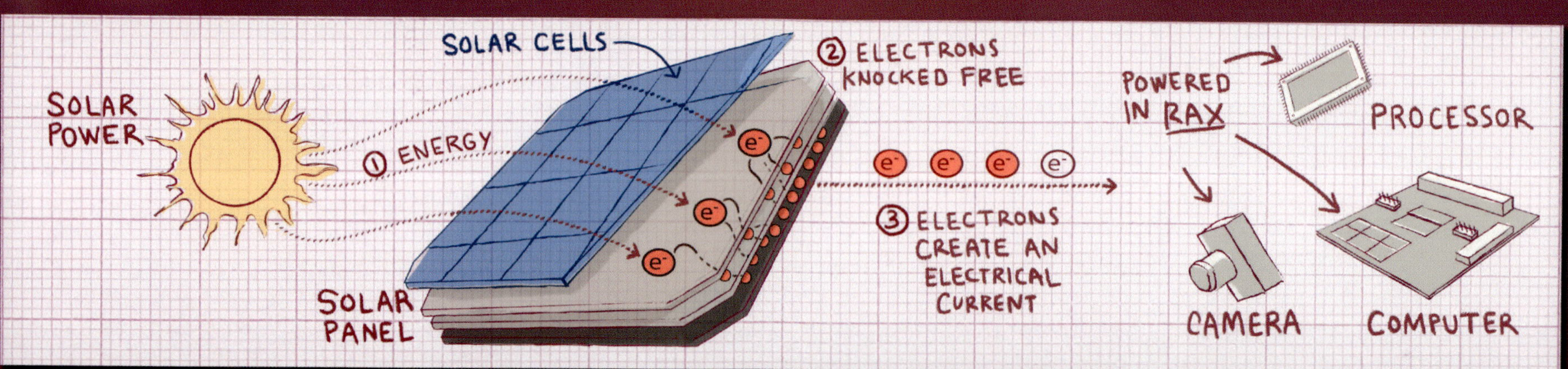

cell phone?

No. There are no cell towers in space to carry the signal.

Email?

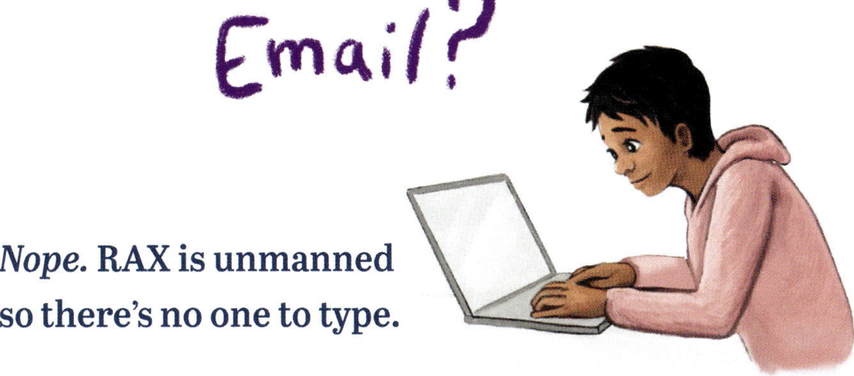

Nope. RAX is unmanned so there's no one to type.

Snail mail?

Negative. Too slow, and you can't drive a delivery truck in space.

What about code?

Yes!

Computer code will allow the team to talk to RAX.

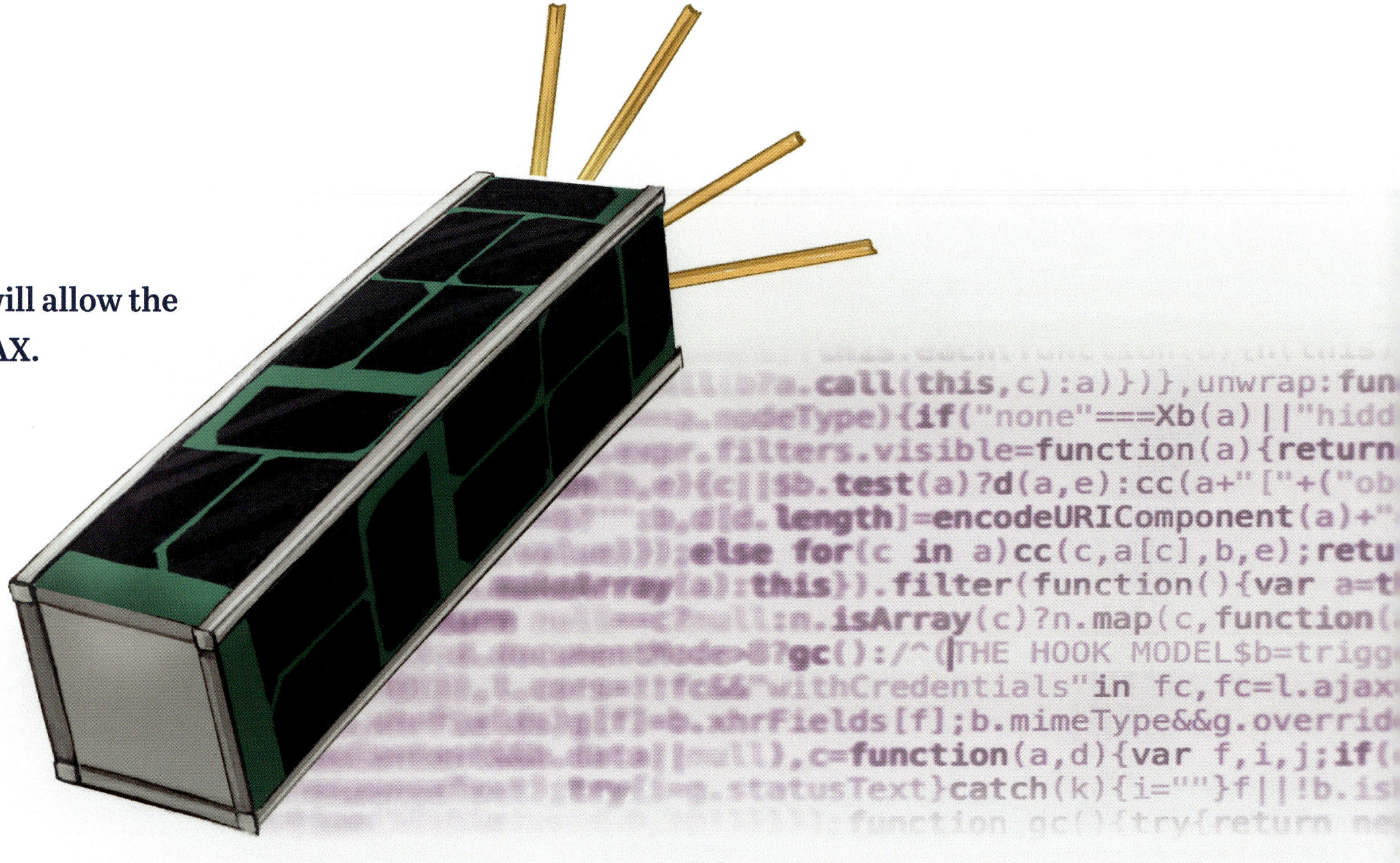

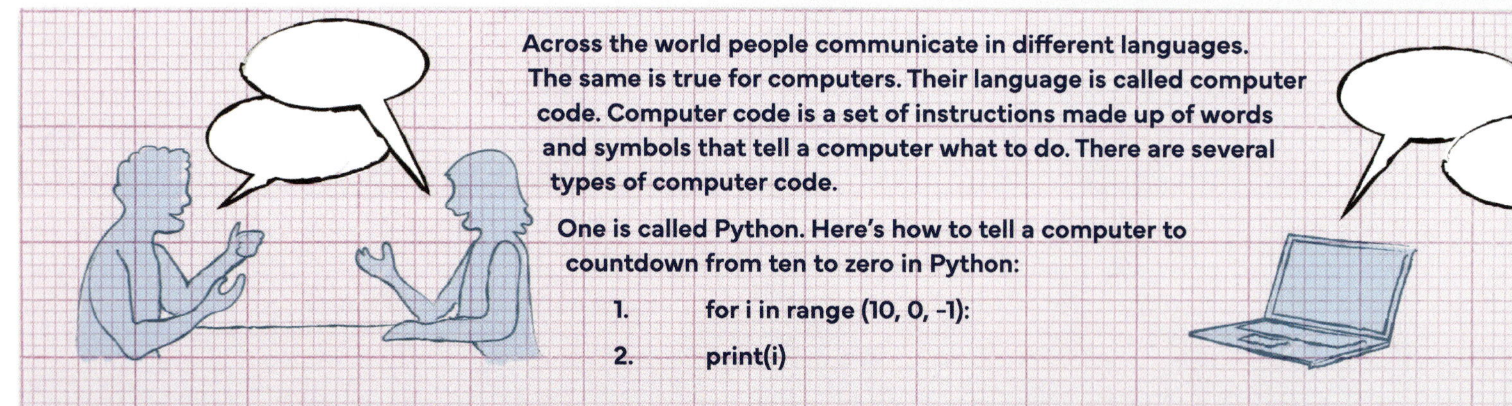

Across the world people communicate in different languages. The same is true for computers. Their language is called computer code. Computer code is a set of instructions made up of words and symbols that tell a computer what to do. There are several types of computer code.

One is called Python. Here's how to tell a computer to countdown from ten to zero in Python:

```
for i in range (10, 0, -1):
print(i)
```

The team designs, builds, and wires a computer. They construct, tweak, and attach four antennas. Now RAX can talk with *blips* and *bleeps!*

But there's still another problem to solve. *How will RAX get into space?*

Balloons?

Nope. They pop too easily.

Airplane?

No. Planes need air for lift, and there's no air in space.

Carrier pigeon?

Yuck. No bird poop on RAX, please.

Could RAX hitch a ride?

We can live on Earth because of the atmosphere that surrounds our planet. Earth's atmosphere is made of mostly oxygen and nitrogen gases. Humans need oxygen to breathe.

As we get farther and farther away from the ground, like hiking up a mountain, the air gets thinner. There is less oxygen, and it is harder to breathe. At about 62 miles (100 km) past Earth's surface there is no more air. This is where outer space begins.

Yes! RAX can use rockets. But traveling by rocket is dangerous. Fierce, fiery blasts! Violent shocks! Blazing heat! Is RAX ready to face these perils?

The team paces and ponders. If RAX fails again, it's back to the drawing board.

Maybe RAX should practice first.

RAX goes to another lab for three tests—Shake, Shock, and Bake.

First, lab workers lift RAX onto a table that jolts, joggles, and jerks like a rocket blasting off the launch pad.

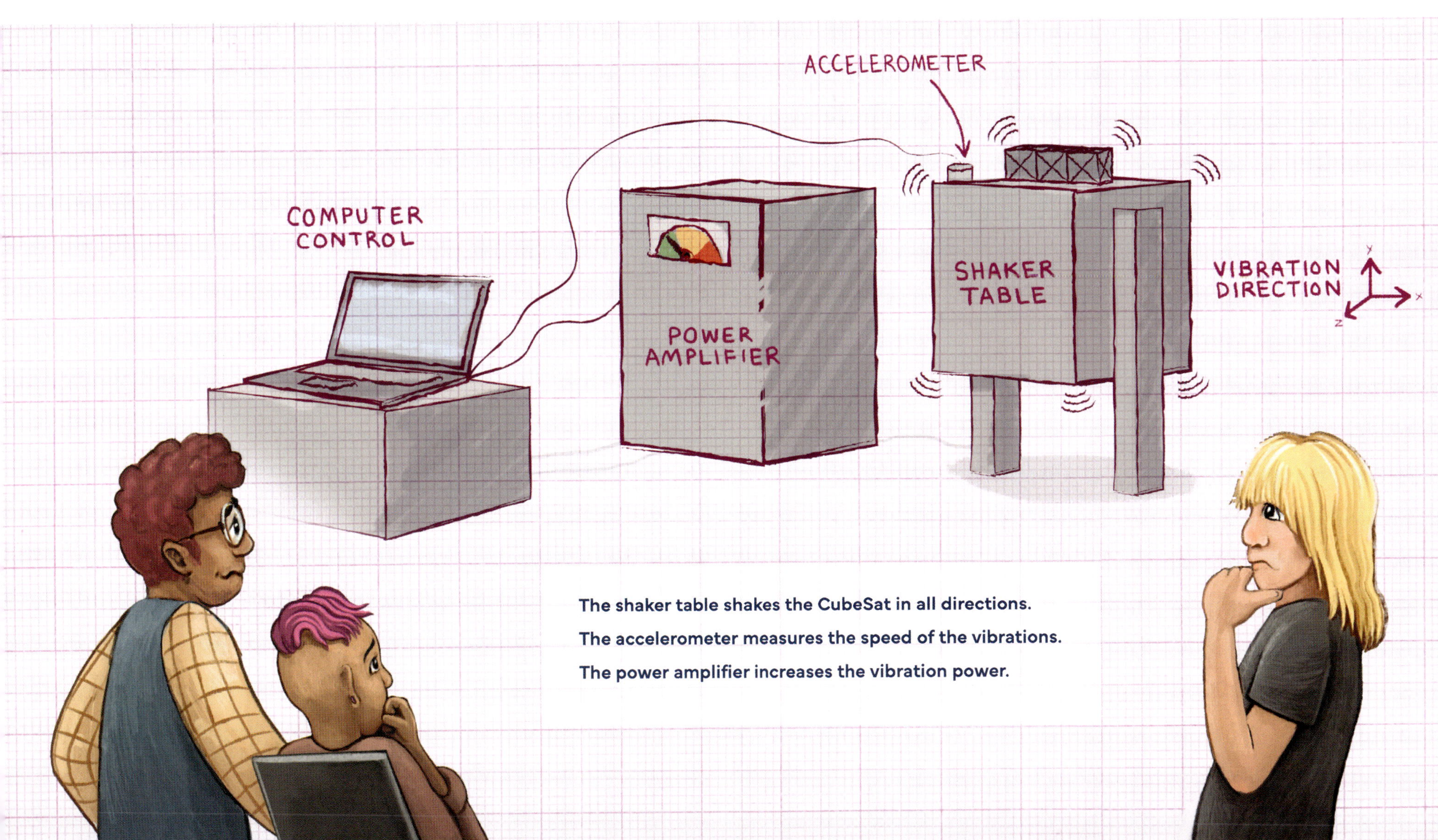

The shaker table shakes the CubeSat in all directions.

The accelerometer measures the speed of the vibrations.

The power amplifier increases the vibration power.

Phew! RAX is shook but is still in one piece. But can RAX hold up during an explosion?

RAX is placed on a metal plate with gun powder.

The lab workers light a fuse,
and...

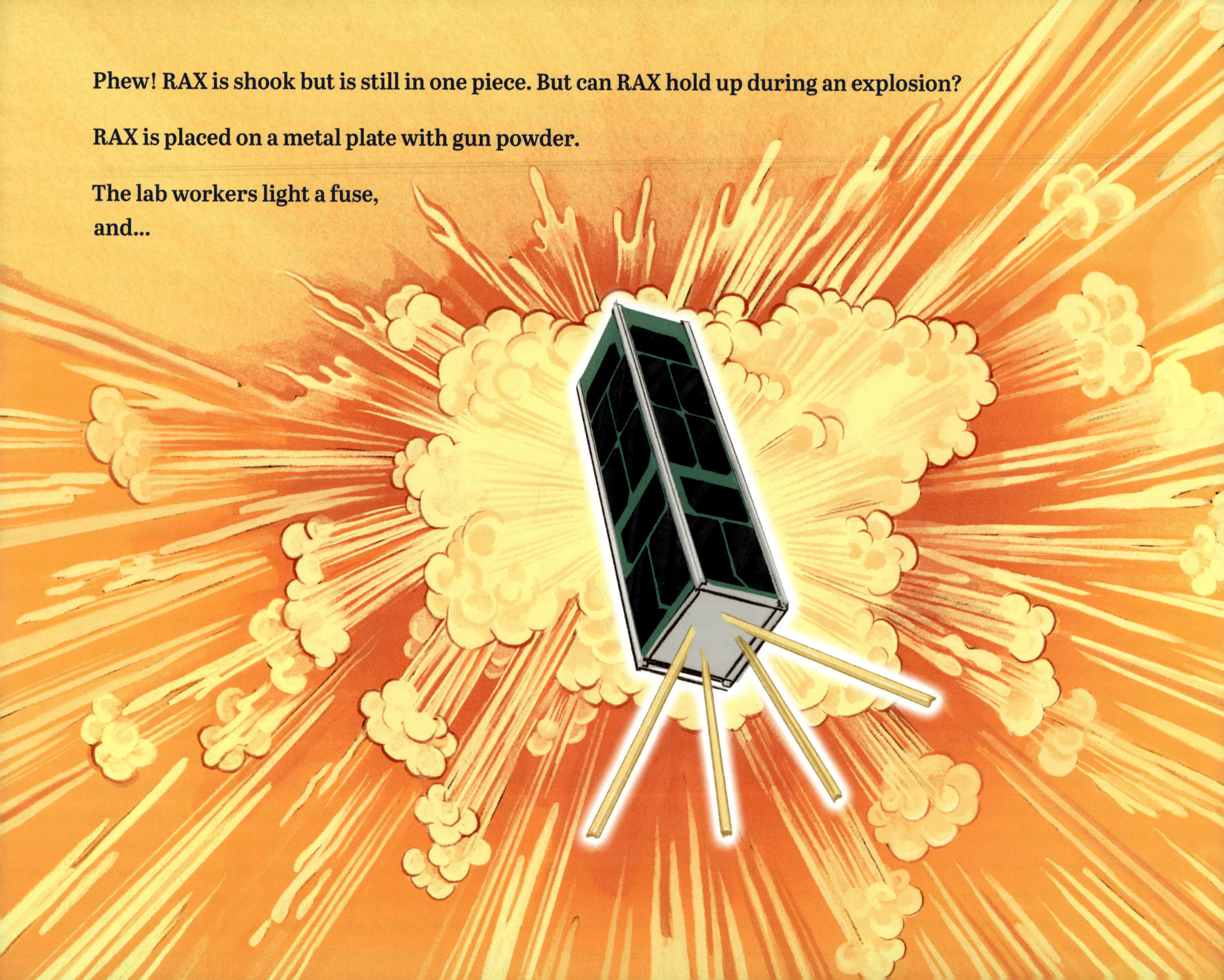

...BANG!

Hooray! RAX survives the violent surge.

SHOCK TEST

The metal plate is where the test item—in this case a CubeSat—is mounted.

Gun powder, when lit, creates an explosion imitating the shock made when one stage of a rocket separates from another during launch.

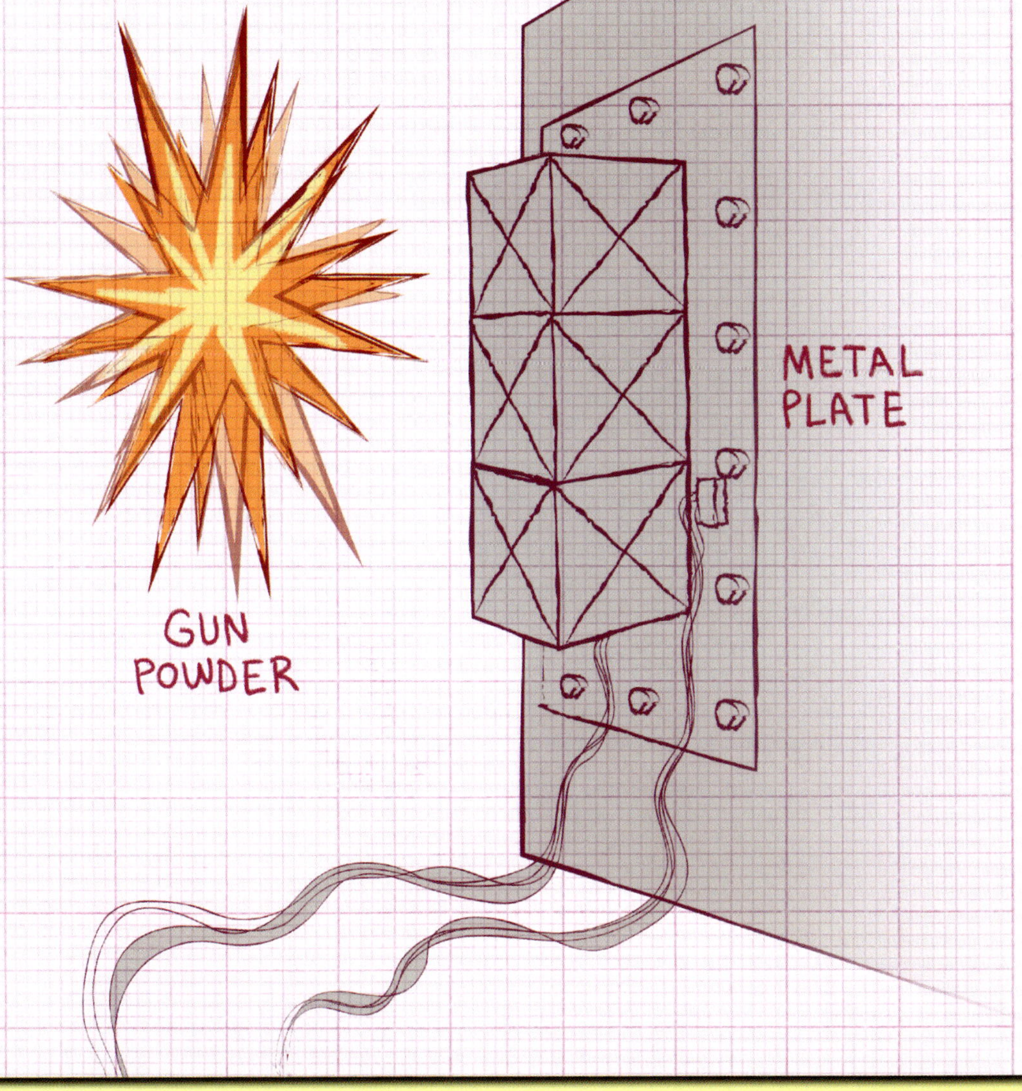

GUN POWDER

METAL PLATE

The final test. Can RAX withstand the sun's baking heat? Lab workers toast RAX like a marshmallow in a tunnel-like oven.

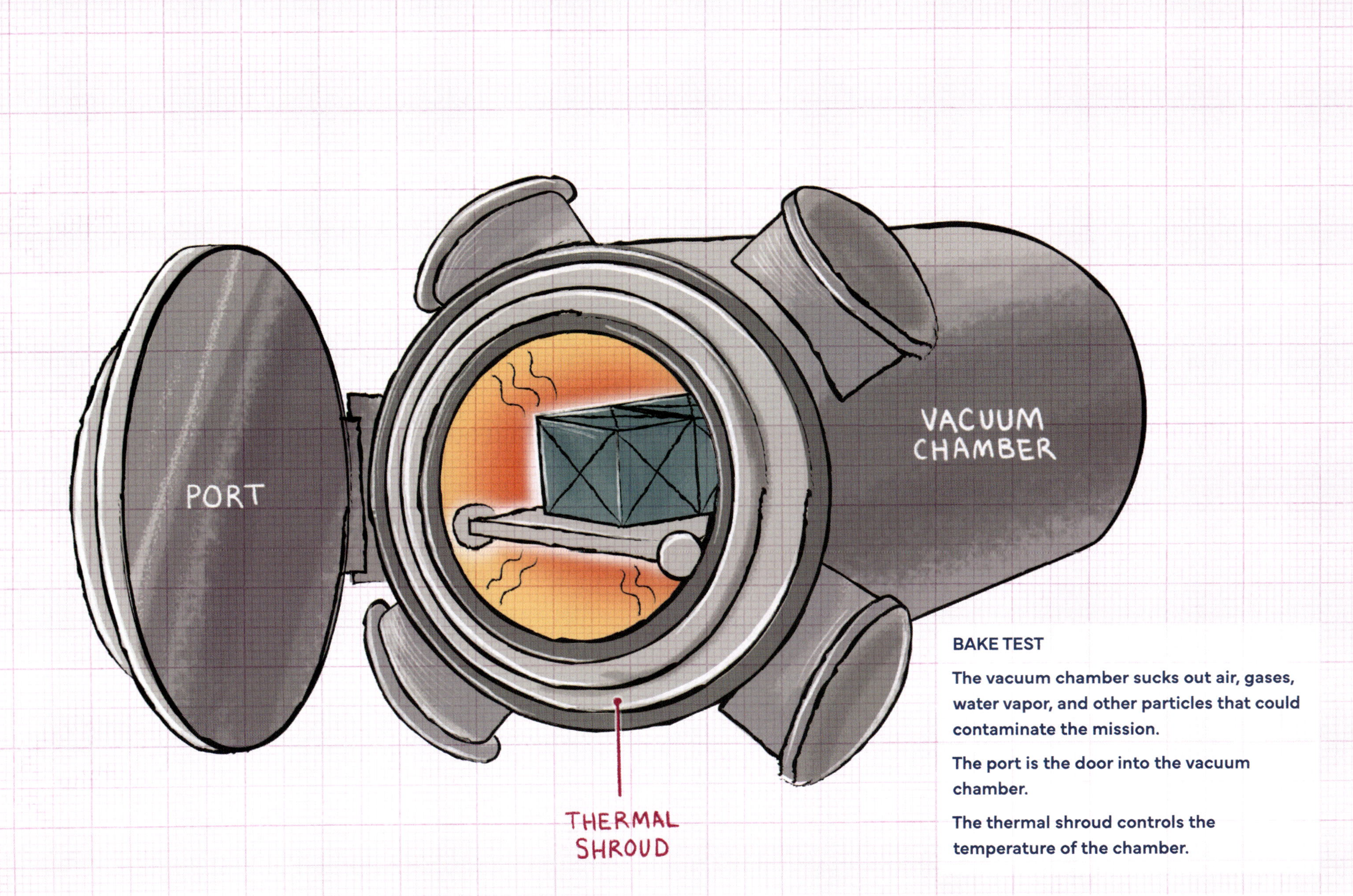

BAKE TEST

The vacuum chamber sucks out air, gases, water vapor, and other particles that could contaminate the mission.

The port is the door into the vacuum chamber.

The thermal shroud controls the temperature of the chamber.

Hot stuff! RAX passes the test without even breaking a sweat! *Next stop?*

The launch site! RAX is wheeled into a clean room. The team must wear head-to-toe "bunny suits" to keep goop, gunk, and grime away from the spacecrafts.

Workers put RAX inside a metal box to keep it safe. The box is packed in the rocket's nose cone. Four other CubeSats and a large NASA satellite are also hitching a ride.

The metal box RAX was placed inside is called a Poly-Picosatellite Orbital Deployer, P-POD for short. P-PODs are made of aluminum and measure 5 x 5 x 16 inches. One can hold up to three CubeSats.

Because RAX was the size of three cube building blocks, it was the only passenger in its P-POD.

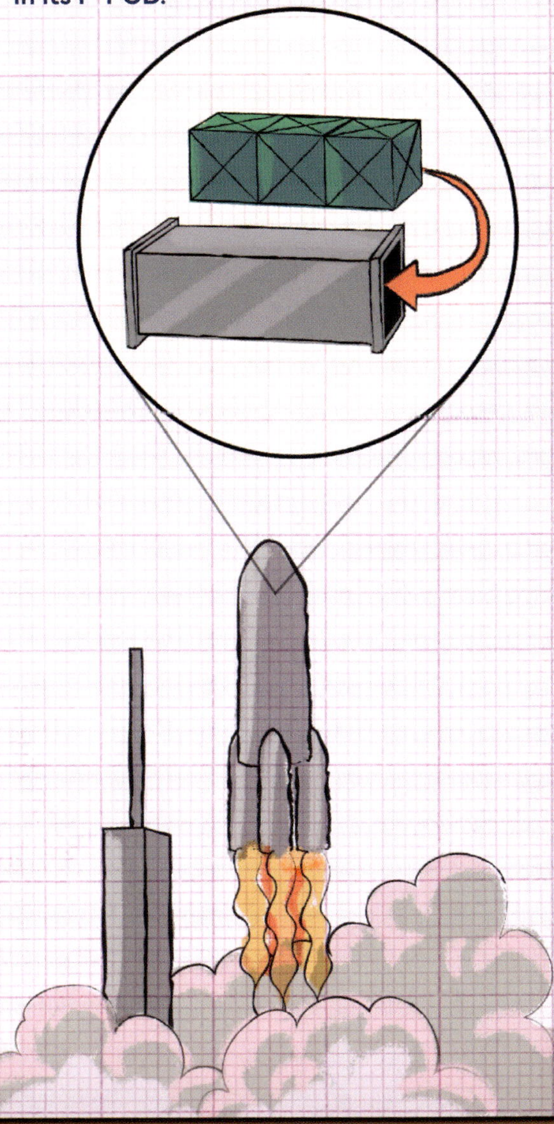

In the control room, the launch team performs the final systems check.

The countdown begins.

Five...four...three...two...one...

LIFTOFF!

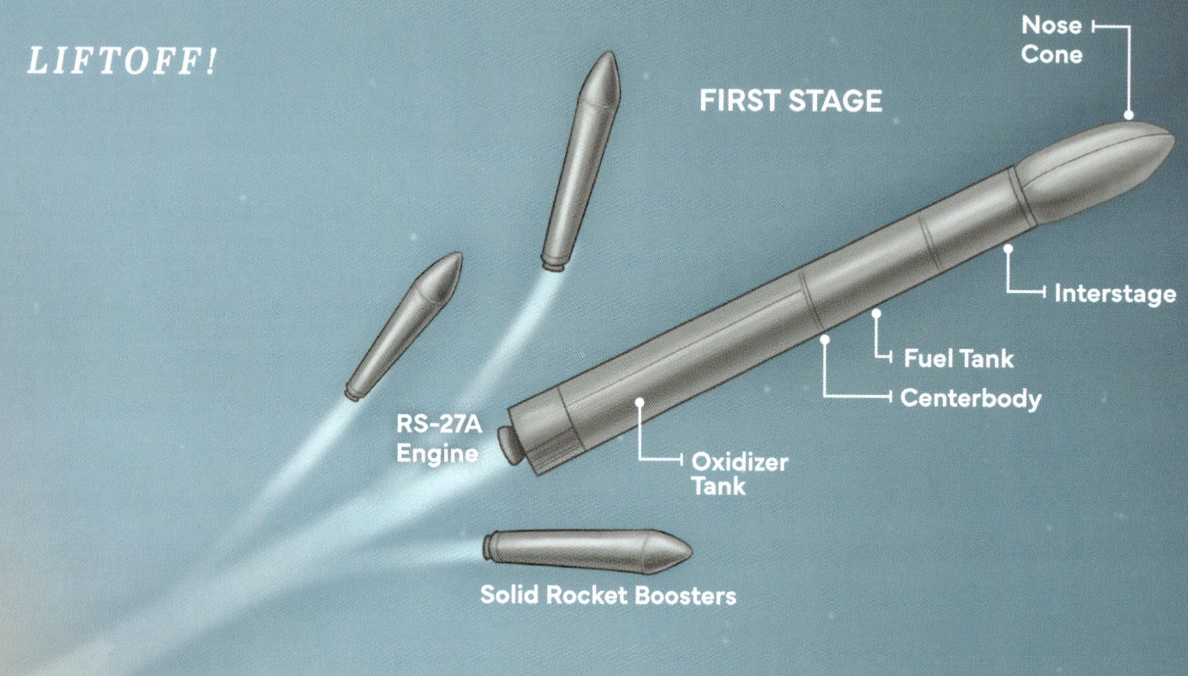

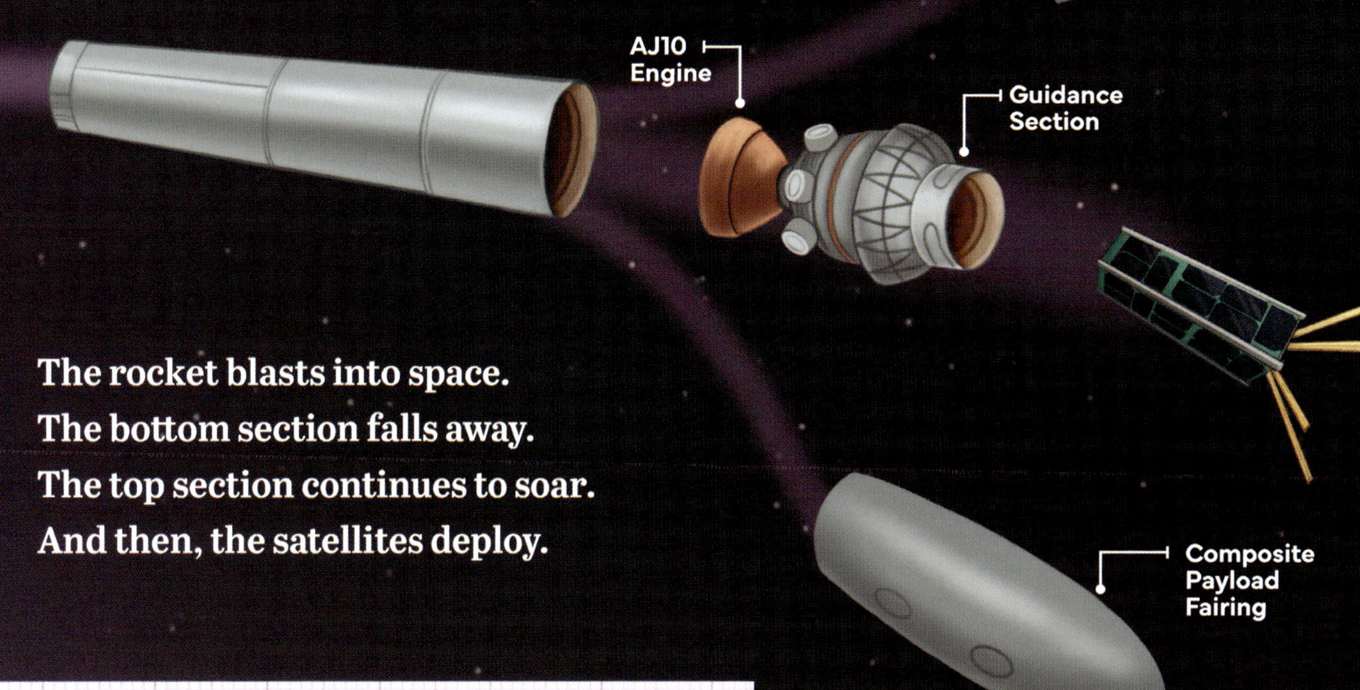

The rocket blasts into space.
The bottom section falls away.
The top section continues to soar.
And then, the satellites deploy.

The FIRST STAGE boosts the rocket upward.

PAYLOAD is the name for the items that go on a rocket to journey into space.

The SECOND STAGE carries the payload to its exit point.

The NOSE CONE is where RAX and the other satellites are housed.

The FUEL TANK holds the fuel needed to launch the rocket.

Everyone's heart races. *Is RAX going to succeed this time?*

NASA's satellite goes first. Another team's CubeSat second. Finally, it's go time!

The metal box holding RAX opens. A spring catapults it into orbit like lava from an erupting volcano.

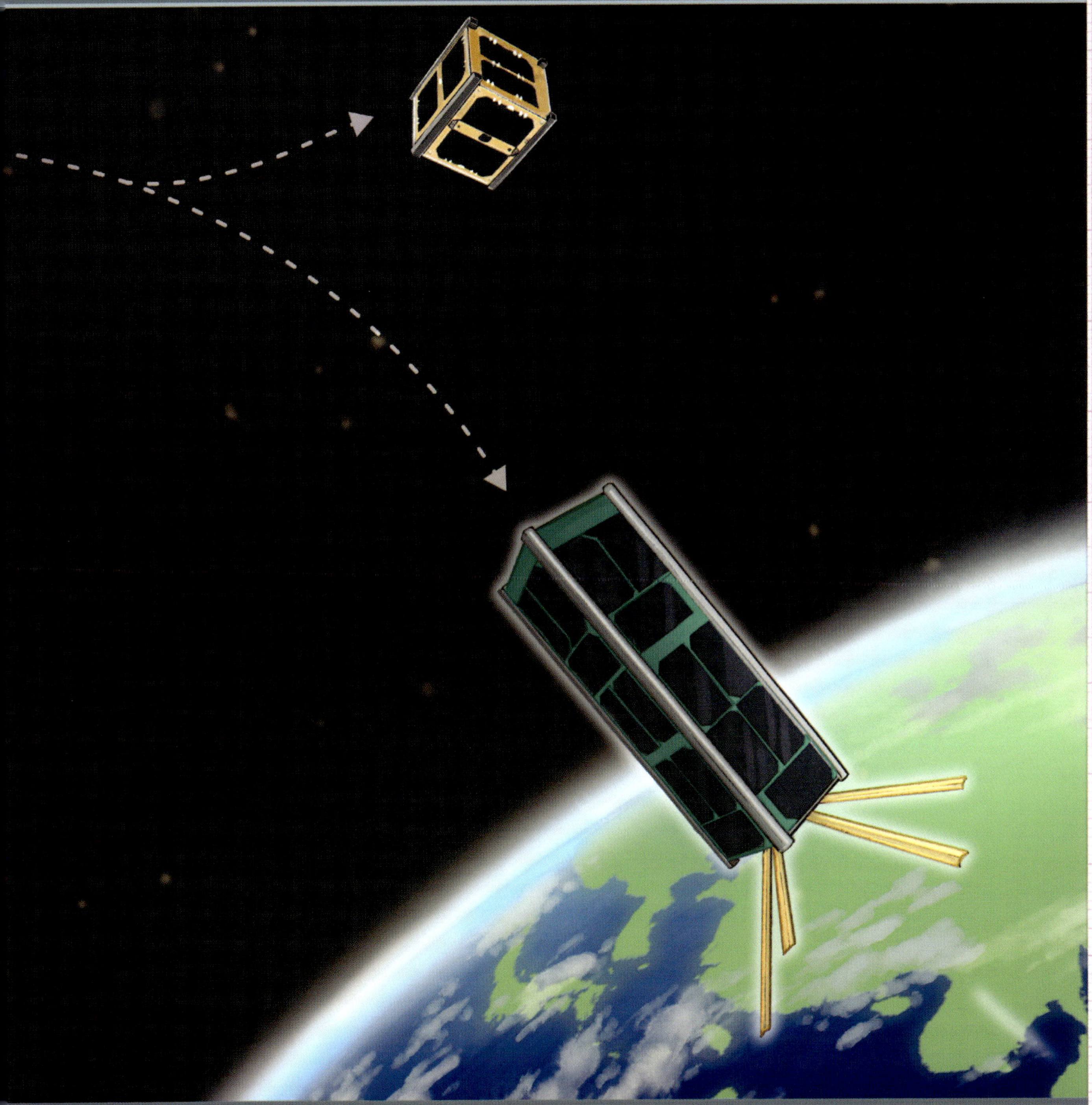

NASA's satellite was called Suomi NPP.

Like the meteorologists who study and report the weather on the news, Suomi's job was to study weather and climate change. Suomi was much larger than RAX. It was the size of a large couch and weighed as much as a female hippo.

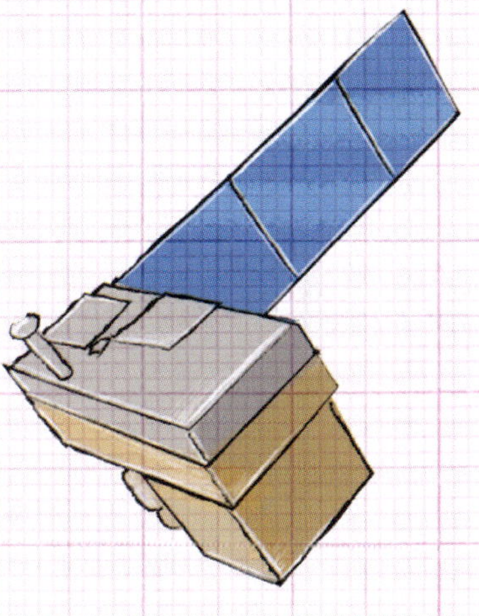

After forty-five minutes in orbit, RAX's solar panels shine as they charge from the Sun's light. Its power supply activates.

Its antennas deploy.

The team waits for RAX to send its first message...

RAX's first *blip* was heard by a ham radio operator in Germany. Ham radios don't need the internet or a cell network. People can operate them anywhere to communicate across states, countries, and even outer space. The eXploration lab hired ham radio operators around the world to listen for RAX's signals as it orbited Earth.

The team hoots, hollers, and cheers!

Their smiles are glowing as bright as RAX's solar panels.

Time to tell RAX to begin its space weather experiments. RAX's computer fires *bleeps* into space. Those *bleeps* bounce off the specks of energy spurting from the sun. RAX photographs the specks and sends the images to the team.

Congratulations! Mission accomplished!

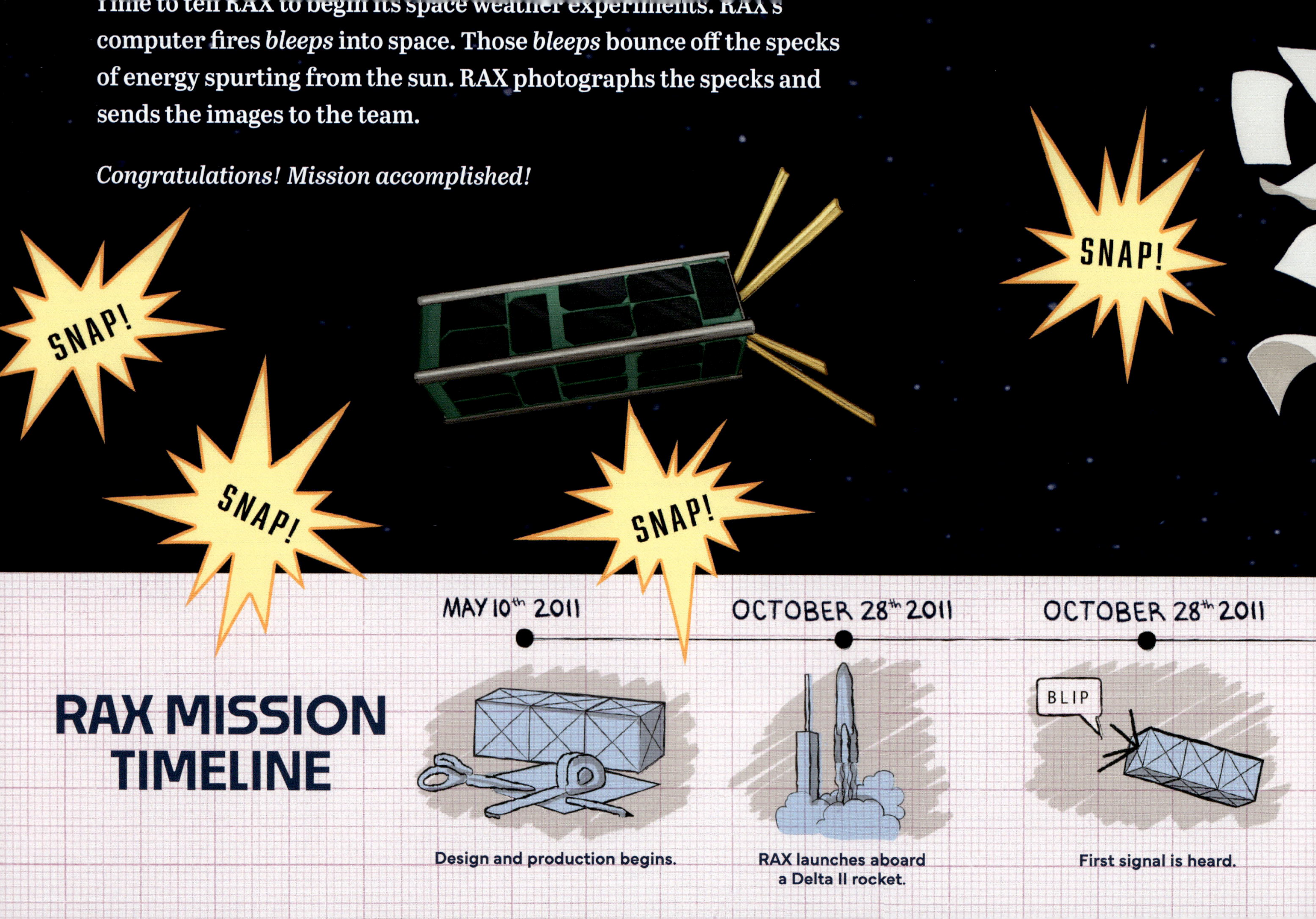

RAX MISSION TIMELINE

MAY 10th 2011
Design and production begins.

OCTOBER 28th 2011
RAX launches aboard a Delta II rocket.

OCTOBER 28th 2011
First signal is heard.

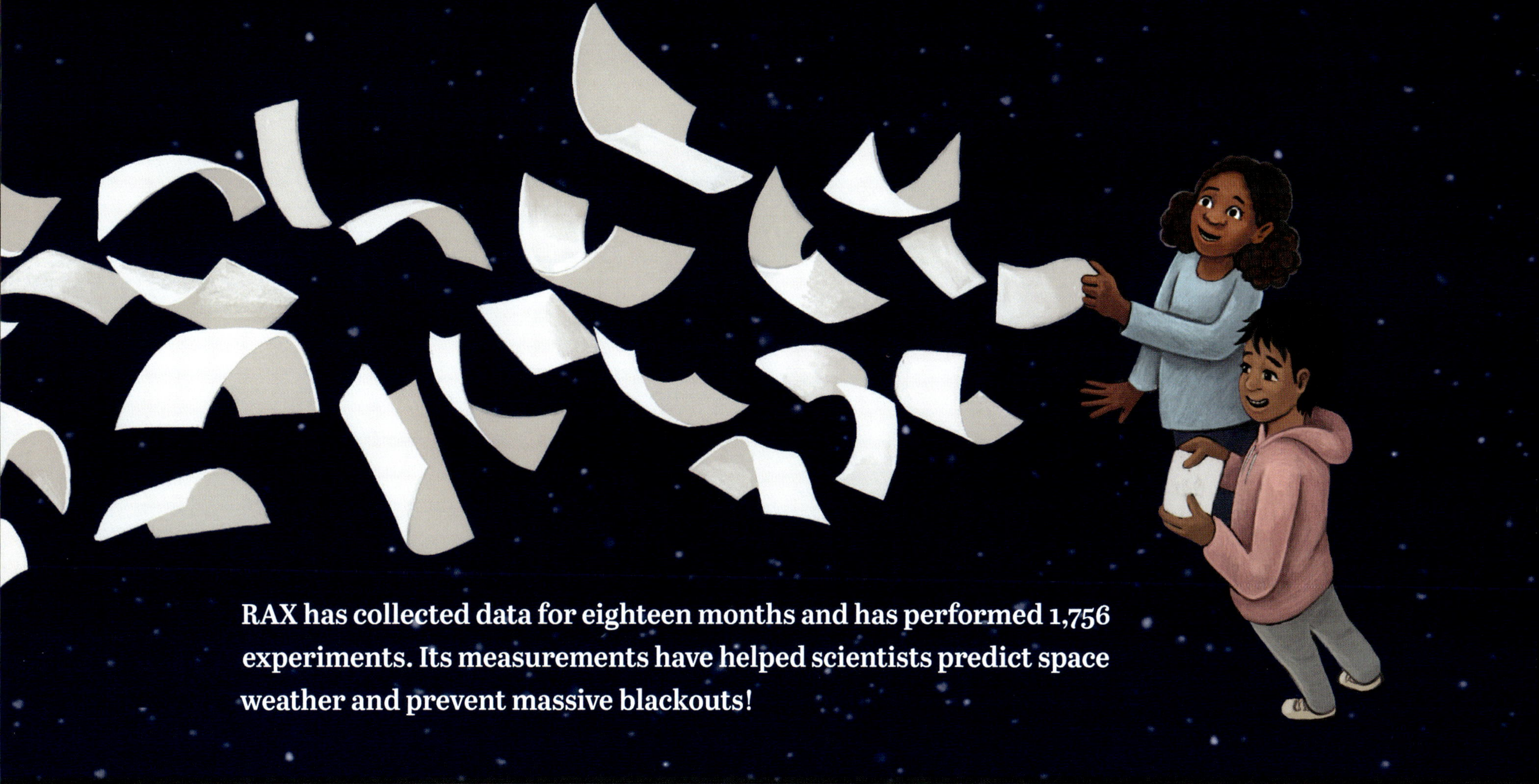

RAX has collected data for eighteen months and has performed 1,756 experiments. Its measurements have helped scientists predict space weather and prevent massive blackouts!

NOVEMBER 22nd 2011

RAX performs its first science experiment.

MAY 17th 2012

RAX has been in orbit for 200 days.

MARCH 11th 2013

RAX has been in orbit for 500 days.

APRIL 22nd 2013

RAX sends its last signal marking the end of the mission.

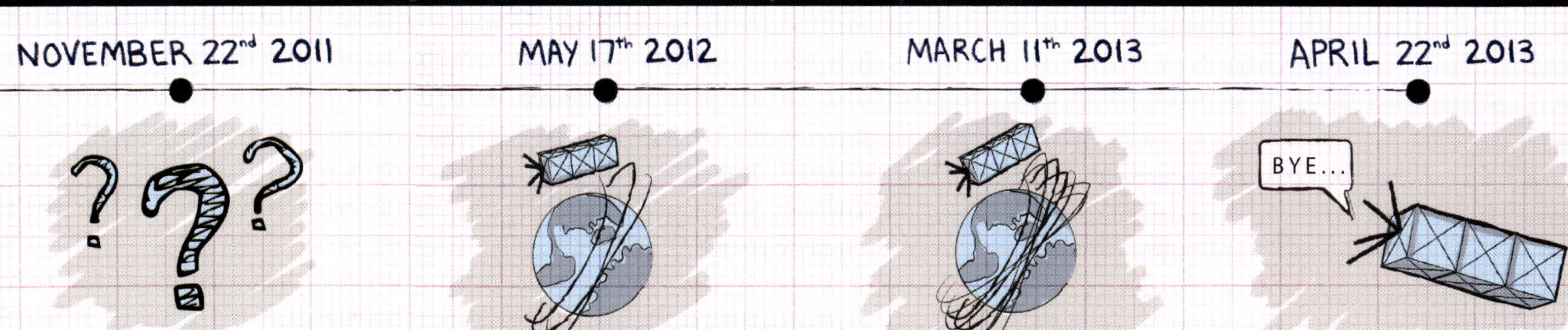

In time, RAX plunges back toward Earth.

It burns up as it re-enters our atmosphere. But don't worry!

Instead, *look to the sky*.

Maybe you will see other mini, but mighty satellites.

Small Satellites... BIG DATA!

Some of the most baffling questions about space can only be answered by large spacecraft, but they are expensive and time-consuming to design and launch. Enter college professors Jordi Puig-Suari (California Polytechnic State University) and Bob Twiggs (Stanford University). Their mission? To develop an inexpensive way for their students to gain hands-on engineering experience with satellites. Inspired in part by the 1990s craze for little stuffed animals known as Beanie Babies, the professors developed the idea of mini satellites—CubeSats.

CubeSats are much cheaper to build and deploy than larger satellites. They require less rocket fuel to lift them into space. And CubeSats can hitch rides with larger payload items making separate launches unnecessary. Low-cost, off-the-shelf parts such as computers, GPS receivers, and antennas save time and labor. CubeSats allow small countries, colleges, and even K-12 schools to participate in satellite development.

The first CubeSats were created by various universities and launched in June 2003. RAX was the first CubeSat designed and built by college students at the University of Michigan eXploration Lab (MXL) under the direction of Professor James Cutler. It was also the first CubeSat chosen for launch by the National Science Foundation, a federal government agency supporting science and engineering.

Today, thousands of CubeSats have been launched into space. Initially, they flew in low Earth orbit, but now engineering teams send them farther. MarCO, another MXL CubeSat, is part of NASA's InSight lander mission to Mars and was the first CubeSat to fly into deep space. Scientists plan to send CubeSats to explore Jupiter and beyond, Because CubeSats provide an immense amount of information about space in a tiny package, they are a valuable way to explore the final frontier.

Bibliography

Bennett, Matt, and Andy Clush. "Michigan Exploration Laboratory (MXL)." Facebook, 2012, www.facebook.com/Michigan.Exploration/.

Cosgrove, Brian. "The Power of The Sun." *Eyewitness Weather*, DK Publishing, 2016, pp. 16–17.

Craddock, Allison, and James Cutler. "Radio Aurora Explorer (RAX-2) and Michigan Multipurpose Minisatellite (M-Cubed): Two CubeSats from the Michigan Nanosatellite Pipeline Launch Together." *Space Times*, 2011, pp. 4–7.

Cutler, James, and Allison Craddock. "The Radio Aurora EXplorer." *Space Times*, 2010, pp. 15–16.

Cutler, James. *The Michigan EXploration Lab*, exploration.engin.umich.edu/blog/.

Dunbar, Brian. "Sun-Earth Connection." *NASA*, NASA, 30 Mar. 2015, www.nasa.gov/mission_pages/themis/auroras/sun_earth_connect.html.

Howell, Elizabeth. "CubeSats: Tiny Payloads, Huge Benefits for Space Research." Space.com, *Space*, 19 June 2018, www.space.com/34324-cubesats.html.

Lipshaw, Suzanne Jacobs. "Professor James Cutler, *Michigan Exploration Lab*." 15 June 2018.

Loff, Sarah. "CubeSats Overview." *NASA*, NASA, 22 July 2015, www.nasa.gov/mission_pages/cubesats/overview.

Mabrouk, Elizabeth. "What Are SmallSats and CubeSats?" *NASA*, NASA, 13 Mar. 2015, www.nasa.gov/content/what-are-smallsats-and-cubesats.

"Space Weather." *National Geographic Kids Ultimate Weather-Pedia: the Most Complete Weather Reference Ever*, by Stephanie Warren Drimmer, *National Geographic*, 2019, pp. 248–258.

"Space Weather." *Ready.gov*, www.ready.gov/space-weather.

"Thunderstorm Basics." *NOAA National Severe Storms Laboratory*, www.nssl.noaa.gov/education/svrwx101/thunderstorms/.

"Tracking the Causes of Space-Based Weather Disruptions." *NSF*, nsf.gov/discoveries/disc_summ.jsp?cntn_id=118950.

"What Is Space Weather?" *NASA*, NASA, 30 Sept. 2019, spaceplace.nasa.gov/spaceweather/en/.

Williams, Matt. "What Are CubeSats?" *Universe Today*, 2 Oct. 2016, www.universetoday.com/82590/cubesat/.

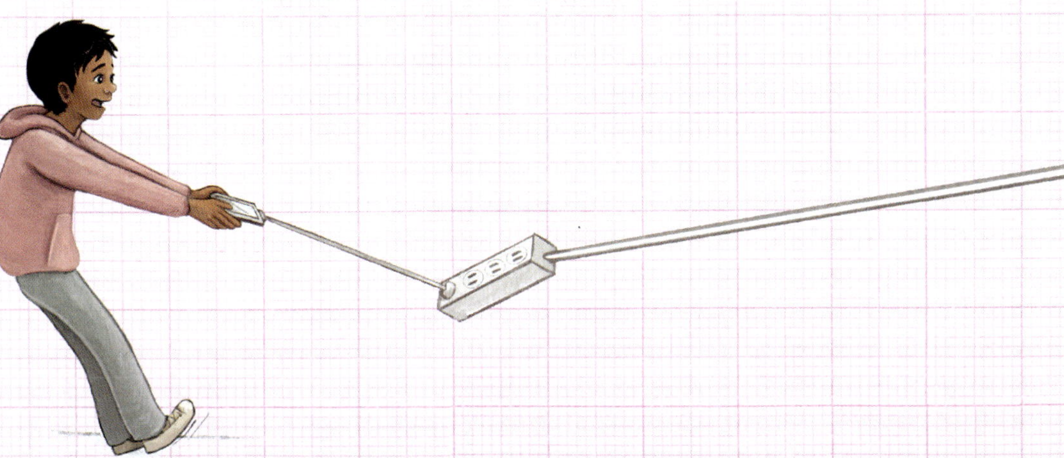

Suzanne Jacobs Lipshaw, Author

Suzanne Jacobs Lipshaw is an award-winning children's book author and former elementary special education teacher passionate about growing young minds, engaging readers, and empowering student leaders. The proud momma of two grown boys, Suzanne lives in Waterford, Michigan, with her husband and furry writing companion Ziggy. When she's not dreaming up new writing projects, you can find her reading, kayaking, hiking, or practicing yoga. To learn more about Suzanne visit her website at suzannejacobslipshaw.com.

Mesa Schumacher, Illustrator

Mesa Schumacher is a science artist and writer who loves learning about just about everything. Her illustrations can be found training doctors and explaining scientific research, her infographics appear in the pages of *National Geographic*, and her animal art is on display throughout the National Zoo in Washington, D.C. Mesa can usually be found living somewhere around the world with her husband and two children—right now they're on the tiny island of Mauritius in the Indian Ocean. Her work can be found at mesaschumacher.com.